MAKE it WORK!

ELECTRICITY

Wendy Baker and Andrew Haslam

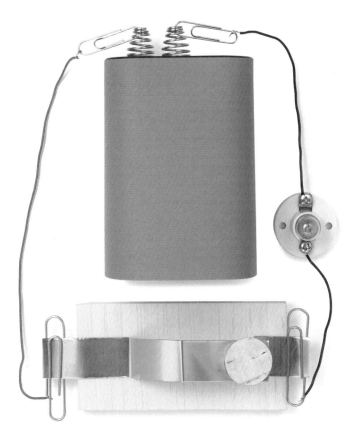

written by
Alexandra Parsons

Photography by Jon Barnes
Consultant: John Chaldecott
Science Consultant: Graham Peacock
Lecturer in Science Education at
Sheffield City Polytechnic Centre
for Science Education

World Book

in association with
ⓦⓒⓝ

MAKE it WORK!
Other titles

Body
Building
Dinosaurs
Earth
Flight
Insects
Machines
Photography
Plants
Ships
Sound
Space
Time

This edition published in the United States in 1997 by
World Book, Inc., 525 W. Monroe, Chicago, IL 60661
in association with Two-Can Publishing Ltd.

© 1997, 1992 Two-Can Publishing Ltd.
Design copyright © 1992 Wendy Baker and Andrew Haslam

**For information on other World Book products,
call 1-800-255-1750, x 2238, or
visit us at our Web site at http://www.worldbook.com**

ISBN: 0-7166-4702-8 (hard cover)
ISBN: 0-7166-4703-6 (soft cover)
Library of Congress Number: 97-60319

Printed in Hong Kong

1 2 3 4 5 6 7 8 9 10 02 01 00 99 98 97

Editor: Mike Hirst
Series consultant: John Chaldecott
Illustrators: Diana Leadbetter and Michael Ogden
Additional photography: John Englefield
Additional editorial and design work: Sharon Nowakowski, Carole Orbell,
Robert Sved and Belinda Webster

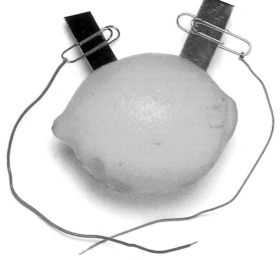

Contents

Being a Scientist 4

Static Electricity 6

Simple Circuits 8

Batteries 10

Lighthouse 12

Complex Circuits 14

Light Bulbs 16

Circuit Game 18

Switches 20

Morse Code 22

Circuit Quiz 24

Magnetism 26

Magnetic Power 28

Magnetic Boats 30

Magnetic Earth 32

Magnetic Fields 34

Electromagnetism 36

Making a Motor 38

Working Motor 40

Electric Train 42

Boosting Power 44

Glossary 46

Index 48

Be Careful! Electricity can be very dangerous. All the activities in this book use batteries as the power source. You should **never** experiment with household electricity or the plugs and sockets in your home.

Words marked in **bold** are explained in the glossary.

4 Being a Scientist

Scientists study the universe and how it works. They ask themselves questions and then work out step-by-step methods to find the answers. Scientists start out with a **hypothesis**, a guess based on known facts. Then they test their idea by doing **experiments**, carefully observing and recording the results.

Investigating electricity is part of the science of **physics**. Physicists study **energy** and **matter** and try to make natural forces work for people. For instance, physicists discovered how to make **hydroelectric** and **atomic** power.

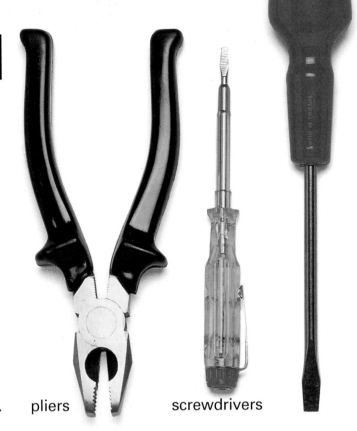

pliers   screwdrivers

MAKE it WORK!

As you do the projects in this book, you will be investigating electricity and the science of physics for yourself. To understand how real scientists work, always use scientific methods. Draw detailed pictures or take photographs of your results, and write down clearly what you have done and observed.

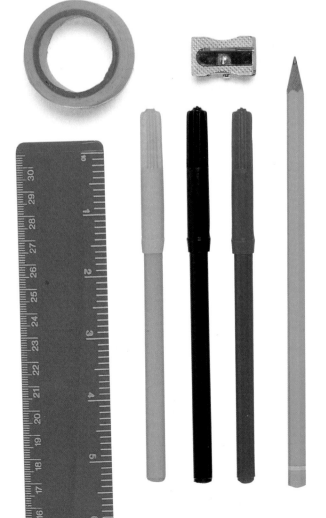

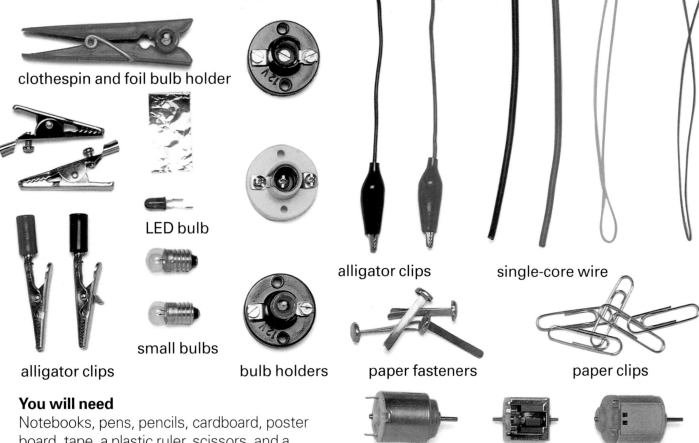

clothespin and foil bulb holder

LED bulb

small bulbs

alligator clips

bulb holders

alligator clips

single-core wire

paper fasteners

paper clips

electric motors

You will need

Notebooks, pens, pencils, cardboard, poster board, tape, a plastic ruler, scissors, and a protractor. Other equipment can be found in hardware and art supply stores.

Wire cutters and **pliers** Special wire cutters are best. You can use an old pair of scissors, but the wire will make them blunt.

Screwdrivers You will need a small electrician's screwdriver with an insulated handle, as well as a larger one to screw pieces of wood together.

horseshoe magnet

6-volt battery

Bulbs and bulb holders Use 6-volt bulbs and matching bulb holders. To make your own bulb holder, wrap aluminum foil around the base of a bulb and hold it in place with a clothespin.

Wire Use single-core, plastic-coated wire.

Clips Alligator clips are found in electrical goods stores, but metal paper clips or paper fasteners are inexpensive substitutes.

Small electric motors Use 3-volt or 6-volt motors from a hobby shop.

Buzzers and **magnets** These are sold in model shops and hardware stores.

Batteries Most of the activities in this book use simple 6-volt batteries.

Be careful! Never touch car batteries and never plug anything into the sockets in your home. Household electricity and large batteries are extremely dangerous!

When you comb your hair, does it sometimes stand straight up and stick to the comb? That's **static electricity**.

All things are made up of tiny particles called **atoms.** When a comb and hair rub together, the outer layers of **electrons** are rubbed off the hair atoms and cling to the atoms of the comb, producing static electricity.

When atoms lose electrons, we say they become positively charged. When they gain electrons, they are negatively charged. Two like charges repel one another—and different charges, like the hair and the comb, attract each other.

Repelling
Rub two balloons against your sweater and tie them to a stick with the rubbed sides facing each other. Because both balloons have the same charge, they swing away from one another.

MAKE it WORK!
With an **electroscope** you can test for the presence and strength of static electricity.

You will need
a glass jar with a plastic lid   bare wire
foil from a candy wrapper   aluminum foil
plastic pen or ruler   piece of silk or wool

Ask an adult to help you push a piece of wire through the lid of a jar. Bend one end of the wire and drape a thin piece of foil from a candy wrapper over it. Crumple a ball of aluminum foil around the other end. Rub a plastic pen with a piece of silk or wool, then hold it over the foil ball. If the pen is charged, the candy wrapper will move. Try rubbing some other objects and see what happens.

Attracting

Make some piranhas like the ones above. Using the graph paper shape at right as a pattern, cut out fish from a single layer of colored tissue paper. Place them on a flat surface. Rub a plastic ruler with a piece of silk or wool to get the ruler's electrons moving. Now pass the ruler over the fish and watch them jump up, attracted by the electric charge.

▼ Make some curly tissue-paper snakes and decorate them using stencils or felt-tip pens. (Be careful because tissue paper tears easily.) Pass a charged ruler over them—and watch them wiggle and wriggle!

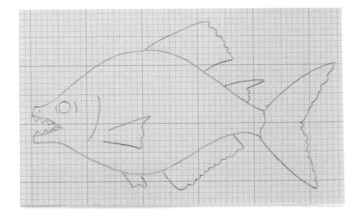

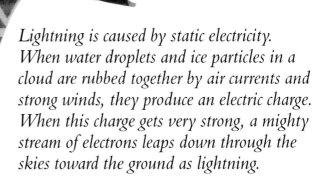

Lightning is caused by static electricity. When water droplets and ice particles in a cloud are rubbed together by air currents and strong winds, they produce an electric charge. When this charge gets very strong, a mighty stream of electrons leaps down through the skies toward the ground as lightning.

8   Simple Circuits

Static electricity itself isn't very useful to us—we have to harness electricity before it can be used. The power that we actually use in our homes is called **current electricity** and is made up of millions of moving electrons.

An electric current is formed when the electrons in a substance, such as a piece of wire, are all made to move in the same direction. To provide us with electrical energy, the electrons must flow in an uninterrupted loop, called an electrical **circuit**.

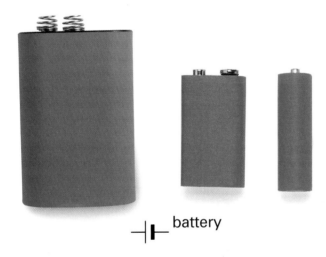

—|⊢ battery

▲ All the projects and activities in this book use small batteries as the source of power.

bulbs   bulb holder

▲ Scientists and electrical engineers use special symbols when they are drawing a diagram of a circuit. This is the symbol for a bulb inside a bulb holder. Did you spot the symbol for a battery above?

MAKE it WORK!

Making a simple circuit is easy, but make sure that all your connections are properly made.

You will need

6-volt battery	bulb and
1 foot of plastic-coated wire	bulb holder

paper clips/alligator clips
pencil, key, cork, eraser, fork, and other household objects

1 Cut the wire in half. Use wire cutters or pliers to strip the plastic coating from the ends, without cutting the wire itself.

2 Wrap the bare end of one piece of wire around each connecting screw on your bulb holder. Tighten the screws to hold the wire in place.

3 Attach the other ends of the wires to clips and attach the clips to the battery **terminals**. If all connections are made properly, the bulb will light.

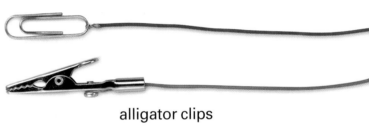

alligator clips

*Anything that an electric current can flow through, such as metal, is called a **conductor**. But materials such as plastic, rubber, and glass do not allow electricity to pass through them. These are called **insulators**. Electrical circuits use both conductors and insulators. The metal in the wire is a conductor that allows the current to flow throughout the circuit. The plastic around the wires and on the bulb holders is an insulator and stops the current from passing into any metal objects that the circuit may be touching.*

▶ All circuits are made up of three basic elements: the conductor (the wire); the **load**, which uses the electricity (in this case a bulb); and the energy source (the battery).

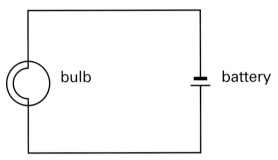

bulb battery

Conductor test

Test some household objects to see if they are conductors or insulators. Make a simple circuit with a gap in it, like the one below right. Touch an object with both wires. If the bulb lights up, you know that electricity must be passing through in order to complete the circuit. That means that the object must be a conductor.

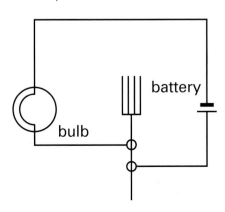

household objects

▶ Electrons flow through some materials better than others. The bulb shines brightly with a good conductor in the circuit and dimly with a poor one.

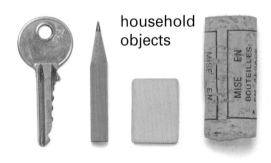

bulb battery

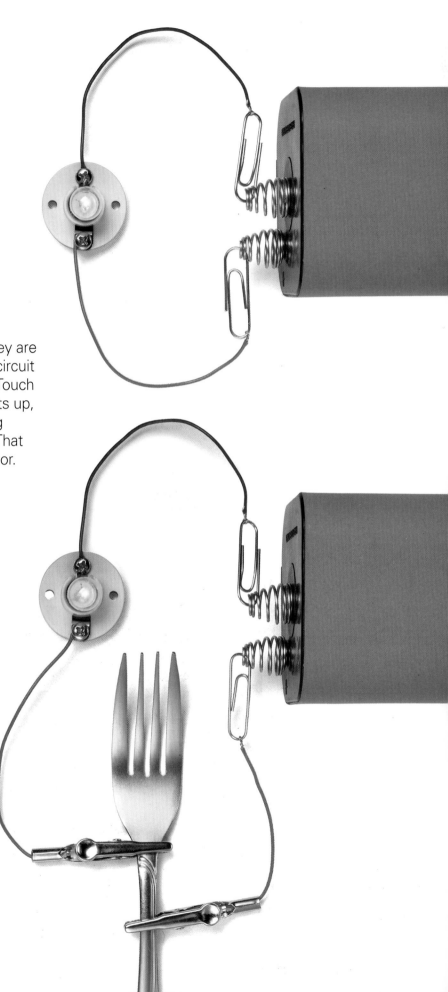

Batteries produce electricity from chemical energy. Usually, two metals, called electrodes, are placed in an **acid** solution called an **electrolyte**. A chemical reaction takes place and creates electric power.

Positive and negative

Our wet battery uses two metals, zinc and copper, to make a current. When these are put into acid, negative electrons move through the liquid from the copper to the zinc. From the zinc, they move down the wire, through the LED, and back up the wire to the copper, completing a circuit and causing an electric flow.

MAKE it WORK!

There are many kinds of batteries. Wet batteries have metal plates in liquid acid. Dry batteries have a chemical paste that separates a carbon rod from a zinc case. Other batteries use other metals and an **alkaline** substance to work.

To make a wet battery you will need

glass jar	white vinegar
wire	alligator clips or paper clips
strip of zinc	piece of copper pipe
light emitting-diode (LED)	

1 Put the strips of metal in the jar and fill it with vinegar. Vinegar is a kind of acid and will be the electrolyte in your battery.

2 When you attach clips and wires as shown, the bulb will light up. However, LEDs work only when wired the right way. If yours doesn't light the first time, reverse the connections.

The first battery was invented in the 1790s by an Italian nobleman, Alessandro Volta. It used disks of silver and zinc, just like our coin battery.

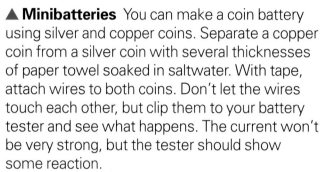

To make a battery tester you will need

balsa wood	cardboard
insulated copper wire	compass
2 screws and washers	wire and clips

1 Secure the compass to the cardboard with copper wire. Attach the ends of the wire to screws on the wooden base.

2 To test a battery, clip wires from the battery terminals to the screws. The compass needle will move. Try this with a brand-new battery, and one that has been used a lot. Can you notice a difference?

▲ **Minibatteries** You can make a coin battery using silver and copper coins. Separate a copper coin from a silver coin with several thicknesses of paper towel soaked in saltwater. With tape, attach wires to both coins. Don't let the wires touch each other, but clip them to your battery tester and see what happens. The current won't be very strong, but the tester should show some reaction.

You can also make a low-voltage minibattery by pushing copper and zinc strips into a lemon.

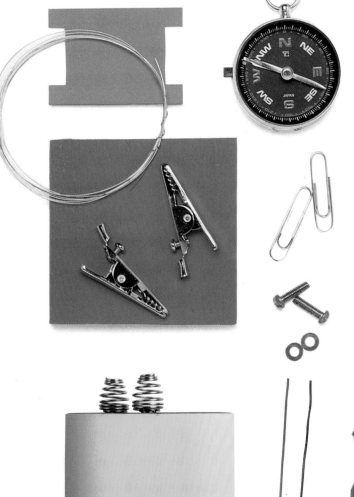

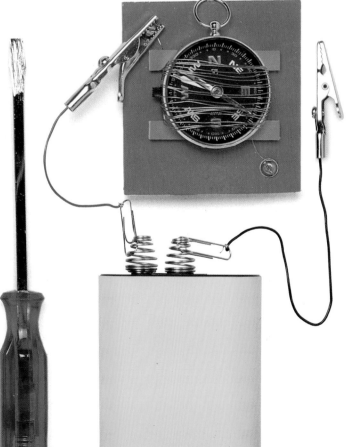

Electricity has many uses—in homes, factories, and schools. It is produced in power stations by burning coal or oil fuels to power electricity **generators**. It can also be produced from nuclear fuel or in hydroelectric turbines.

MAKE it WORK!

Lighthouses were among the first users of electric power. Put your circuit-building know-how to good use and make a battery-operated minilighthouse for your bedroom.

You will need

thin cardboard	glue and tape
utility knife/scissors	wire
bulb and bulb holder	6-volt battery
alligator clips/paper clips	toothpick

1 Make a round tube from a piece of white cardboard and decorate it with red stripes. You could also use the tube from a roll of toilet paper and cover it with white paper.

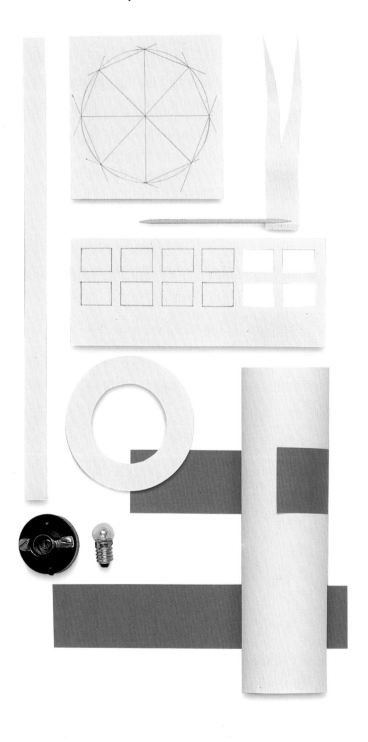

2 To make the balcony, cut out a circle of cardboard, make a hole in the center, and glue or tape it to the top of the tube. Glue a strip of cardboard around the edge of the balcony to make the railing.

3 Attach two long wires to a bulb holder and tape the bulb holder into place at the top of the tube. Push the wires down through the tube and out the bottom.

4 Use a strip of cardboard to make the windows at the top of the lighthouse. Using the picture on the left as a guide, cut out small squares with scissors or ask an adult to use a utility knife. Then bend the cardboard to make a cylindrical shape and glue it in place on the balcony.

5 To make the roof, draw a circle and make a cut from the rim to the center. Fold the circle to form a cone and glue it. Make a flag from paper and a toothpick.

6 Attach the ends of the wires to a battery and the bulb in your lighthouse will light.

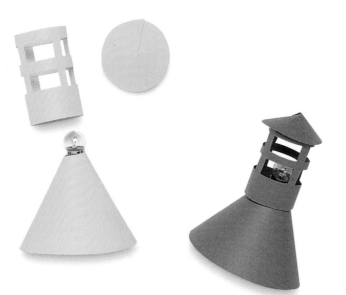

The ancient Egyptians were probably the first people to build lighthouses. They began by lighting bonfires on hilltops to guide their ships. During the third century B.C., they built the tallest lighthouse ever, the Pharos of Alexandria, which was over 400 feet (122 meters) high!

*Scientists measure electricity with two separate units called volts and watts. **Volts** measure electrical force, the amount of power produced by a source of electricity, such as a battery. **Watts** measure the electrical power at the point where it is actually used—in an electric heater or light bulb for instance.*

▲ To hide the battery, make a "rock" out of pieces of old cardboard, stuck together in a jagged shape and painted. Make a series of buoys like those on the next page to go around your lighthouse.

All the parts of an electrical circuit must be joined up to one another so that the current can flow. There are two basic ways to wire a circuit with more than one **component** (or part)—in series or in parallel.

MAKE it WORK!

In a series circuit, the electric current flows along a single path, going through each component in turn. If one component is removed or breaks (when a Christmas-tree bulb burns out, for instance), all the other components will stop working, too.

In a parallel circuit, each component is connected to the power source on its own branch of the main circuit. Even if one of the bulbs in a parallel circuit burns out, the other bulbs will continue to shine because their own branches of the circuit remain complete.

You will need

thin cardboard or construction paper
scissors/utility knife   glue and tape
batteries   wire
bulbs and bulb holders
alligator clips/paper clips

1 You are going to make a string of buoys like the ones used to mark shipping lanes. For each buoy you will need to cut out the shapes you see below from thin cardboard: a semicircle for the body, a strip with windows for the lantern, and a circle with a slit in it for the cone-shaped top. Ask an adult to cut out the windows with a utility knife.

2 Assemble the buoys as shown below, taping a bulb holder firmly into the body of each buoy.

3 Wire the series circuit as shown on the left-hand side of the opposite page. Run a wire from bulb holder to bulb holder, completing the circuit from the last buoy back to the battery.

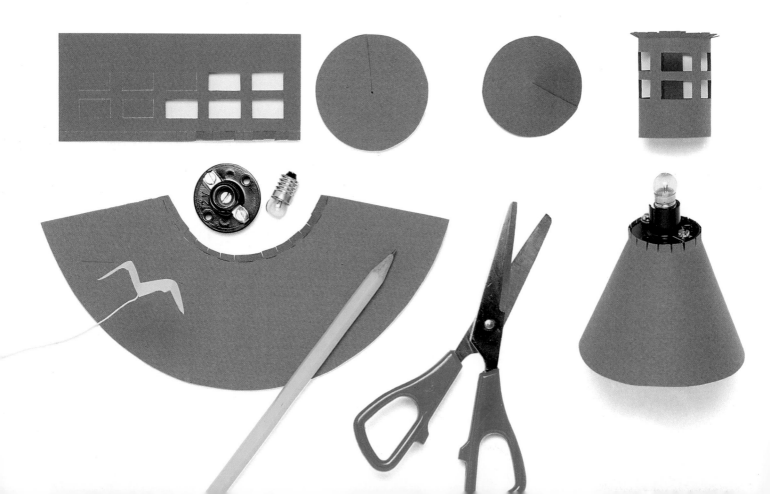

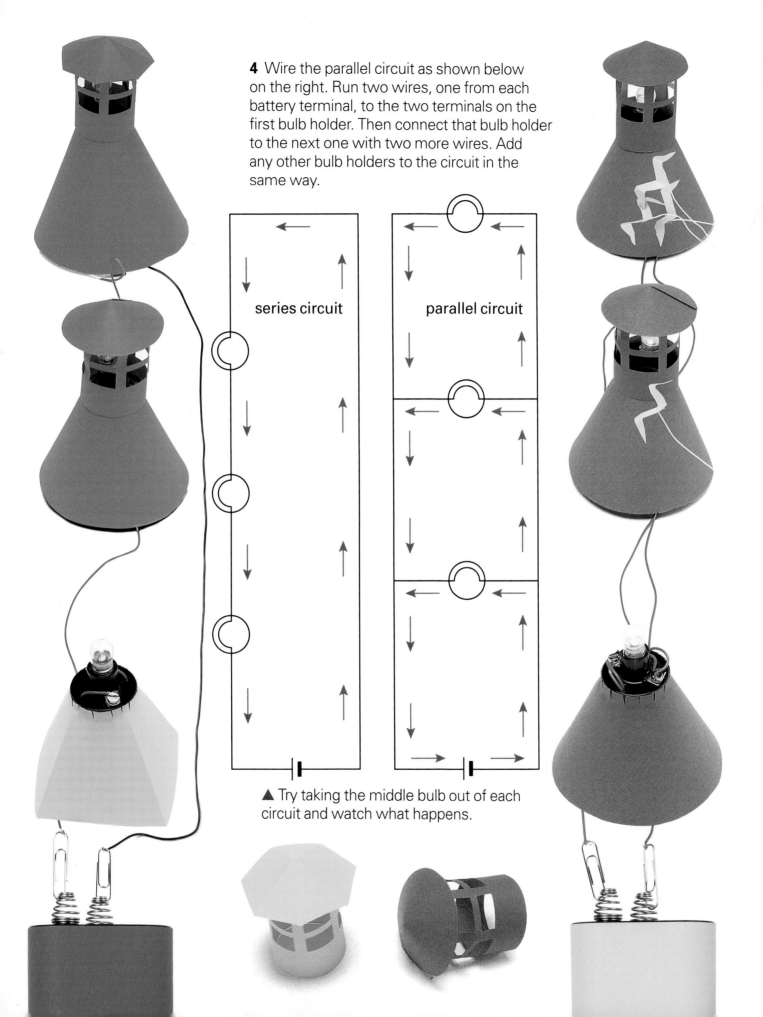

4 Wire the parallel circuit as shown below on the right. Run two wires, one from each battery terminal, to the two terminals on the first bulb holder. Then connect that bulb holder to the next one with two more wires. Add any other bulb holders to the circuit in the same way.

series circuit

parallel circuit

▲ Try taking the middle bulb out of each circuit and watch what happens.

16 Light Bulbs

Light bulbs are used to produce light from electricity. Each bulb contains a thin metal thread called a **filament**. When an electric current forces its way through this thin part of the circuit, the filament glows a bright white color and the bulb gives off light.

You will need
old light bulbs
scissors
glue or tape
cardboard
ruler

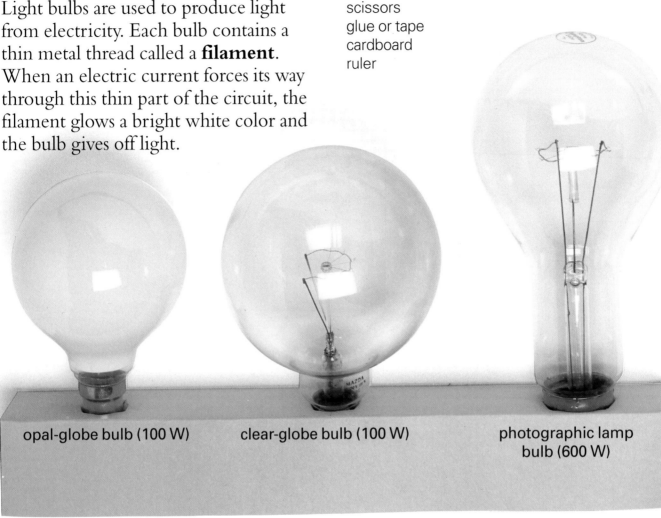

opal-globe bulb (100 W) clear-globe bulb (100 W) photographic lamp bulb (600 W)

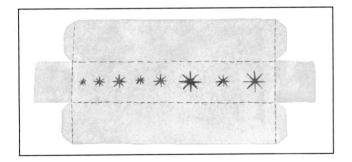

MAKE it WORK!

Light bulbs come in all shapes and sizes. Collect different light bulbs and make a special box to display and store your collection. New bulbs are expensive, so just collect ones that have burned out. Always handle light bulbs carefully because the glass is very delicate.

1 Decide what size you want your box to be. Then cut out a flat shape like the one on the left.

2 Fold the cardboard up along the dotted lines. Tuck in the corner flaps and glue or tape them in place.

3 Carefully cut star-shaped slits in the top of the box. Press the ends of the bulbs into the centers of the stars. Make sure that the bulbs fit firmly.

4 Label your collection. Mark whether each bulb has a filament or a fluorescent tube and also how bright it is. You can tell the brightness of a bulb from the number of watts (W) of power that it uses.

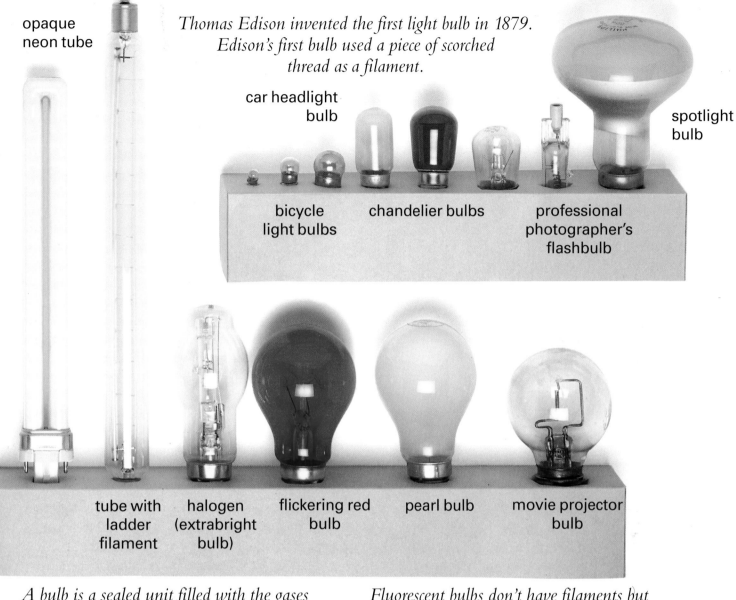

opaque
neon tube

*Thomas Edison invented the first light bulb in 1879.
Edison's first bulb used a piece of scorched
thread as a filament.*

car headlight
bulb

spotlight
bulb

bicycle
light bulbs

chandelier bulbs

professional
photographer's
flashbulb

tube with
ladder
filament

halogen
(extrabright
bulb)

flickering red
bulb

pearl bulb

movie projector
bulb

*A bulb is a sealed unit filled with the gases
nitrogen or argon. It contains no oxygen, the
gas in the air that substances need to burn.
Nitrogen or argon lets the filament glow
but doesn't allow it to catch fire.*

*Fluorescent bulbs don't have filaments but
contain a gas under pressure. When electricity
passes through the bulb, the gas gives off light,
but the bulb stays cool.*

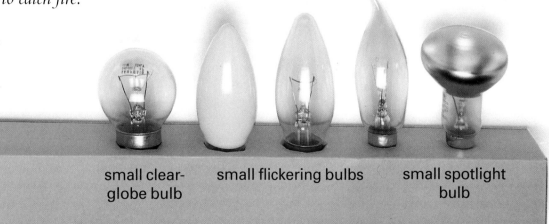

small clear-
globe bulb

small flickering bulbs

small spotlight
bulb

18 Circuit Game

An electric current must always flow through a complete circuit. Current can't flow in a broken circuit because electrons have to keep moving in a continuous stream.

MAKE it WORK!

How steady is your hand? Find out with this circuit game. At the start of the game, the circuit is broken, so the light is off. If your hand shakes as you move the playing stick along, the loop touches the wire, the circuit is completed, and the bulb lights up!

You will need

battery	wire
bulb and bulb holder	dowel
alligator clips/paper clips	colored tape
wire coat hanger	sandpaper
screw eyes, large and small	
balsa wood and wood glue, or shoe box	

Some coat hangers are coated with a thin layer of clear plastic so they won't mark your clothes. Remove the plastic with a piece of sandpaper—otherwise the plastic will insulate the wire and the circuit won't work.

1 Make the playing stick by screwing a large screw eye into the end of the dowel. Wrap one end of a long piece of stripped wire around the base of the eye. Tape the wire along the dowel. Some wire should be left hanging at the end.

2 Make a box by gluing together pieces of wood, or use a shoe box. Paint the box, then divide into sections with colored tape.

3 Position the bulb holder at one end of the box, wire it, and push the wires through the top of the box.

screw eye dowel

wire

4 Twist a small screw eye into each end of the box. If you are using a shoe box, you may have to tape them in place. Bend and twist the coat hanger wire to make the top part of the game. Thread the bent wire through the screw eye in the playing stick. Then connect the ends of the coat hanger wire to the screw eyes on the box.

5 Beneath the box, connect one of the wires from the bulb holder to the battery. Put an alligator clip on the other bulb holder wire and attach it to one end of the bent coat hanger.

6 Connect the wire from the playing stick to the free battery terminal. Now you're ready to play!

circuit diagram of the circuit game

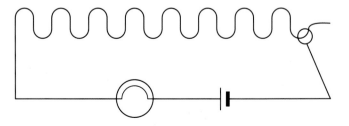

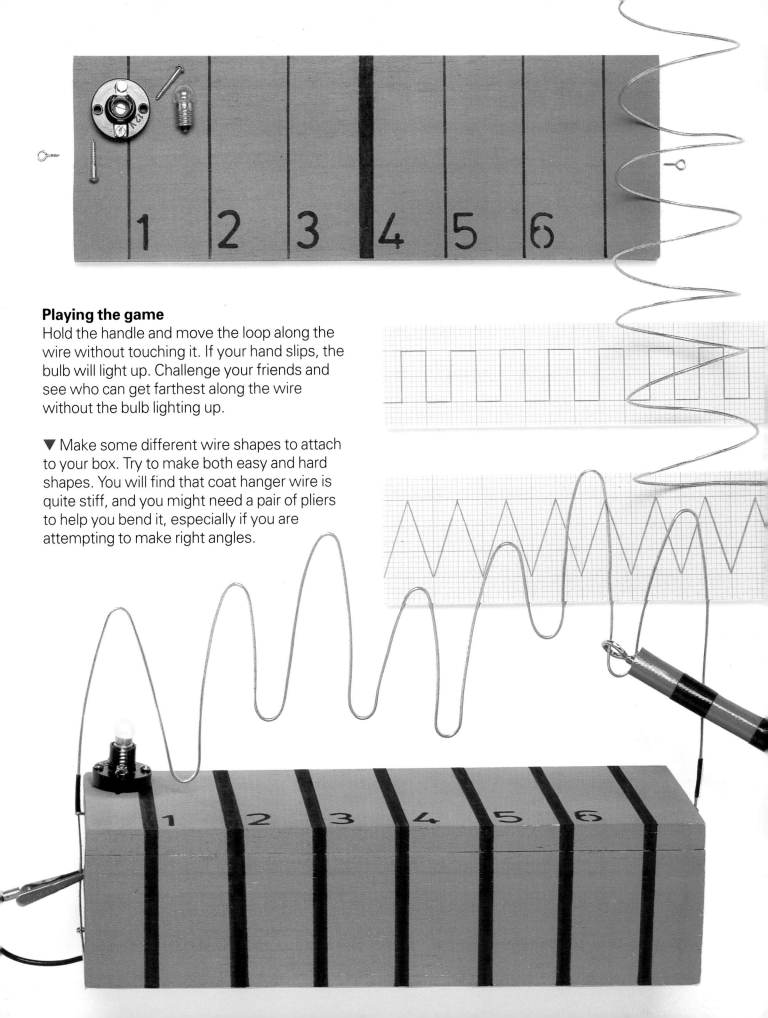

Playing the game

Hold the handle and move the loop along the wire without touching it. If your hand slips, the bulb will light up. Challenge your friends and see who can get farthest along the wire without the bulb lighting up.

▼ Make some different wire shapes to attach to your box. Try to make both easy and hard shapes. You will find that coat hanger wire is quite stiff, and you might need a pair of pliers to help you bend it, especially if you are attempting to make right angles.

Switches are used to turn electrical circuits on and off. When they are switched off, they break the circuit so that electricity can't flow around it. When they are switched on, they complete the circuit, allowing the electricity to flow through.

MAKE it WORK!

Switches can be made to work in lots of different ways. For instance, you may not want a light to go out completely but just to be a little less bright. Or you may need a switch that can turn a buzzer on and off very quickly to make a special pattern of signals. Here are four different types of switches for you to try.

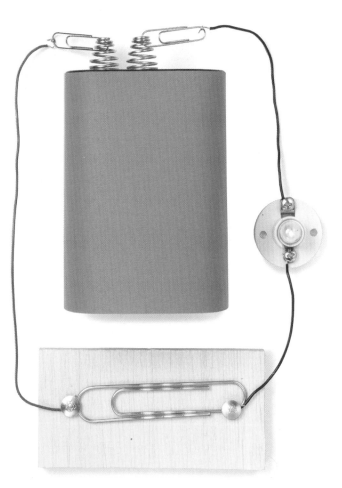

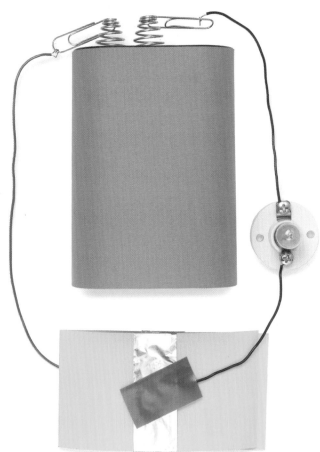

Simple switch

This is a simple on/off switch. When it is on, the current flows through the circuit; when it is off, the current stops. Wire a simple circuit, like the one at the top of page 9, but leave a break in the wires. Make a switch as shown above, using a block of balsa wood, a paper clip, and two metal thumbtacks. When the clip touches both thumbtacks, the switch is on.

Pressure switch

This type of switch can be used to make a doorbell ring when someone steps on a doormat. Wire a circuit as before. Fold a piece of cardboard in half. Wrap strips of foil around each half of the cardboard so that they touch when pressed together. Tape the wires to the foil on the outside of each side of the cardboard. When the strips of foil touch, the switch is on.

You will need

6-volt batteries
bulb holders
balsa wood
paper clips
thumbtacks
aluminum foil
thin copper strips

cork
wire
tape
bulbs
pencil
cardboard
utility knife
alligator clips

◀ These are low-voltage switches from a model shop. You can include them in any of the circuits shown in this book.

Dimmer switch

Electricity can pass through the **graphite** in a pencil, but it is hard work. Graphite is called a **resistor**, because it offers resistance to the electric current. You can use a graphite pencil resistor to make a dimmer switch. The longer your pencil lead, the more resistance there is and the dimmer your light will be.

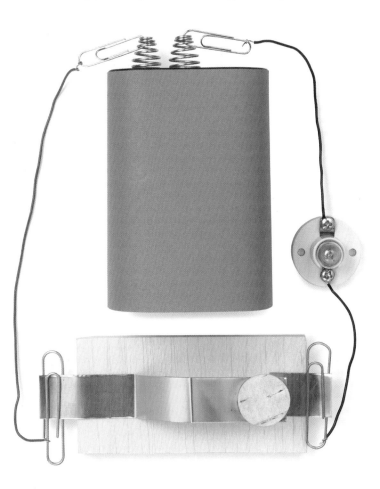

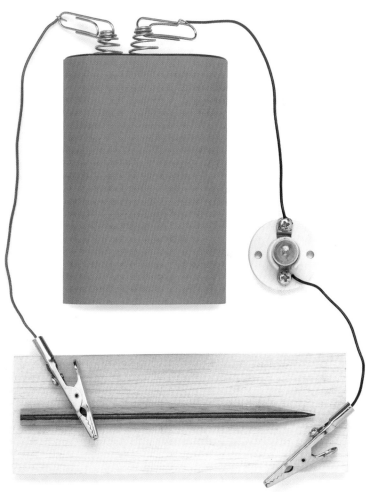

Tapper switch

This switch is used by Morse code operators. It gives the operator total control over the length of time the circuit is complete or broken. The switch is on when the two strips of copper are pressed together. It returns automatically to the "off" position when not in use. The full instructions for how to make a Morse code tapper are given on the next page.

Make a simple circuit as before, but fit alligator clips to the free ends of the wire. Soak a pencil in water, and then ask an adult to slice it open down the middle with a utility knife. (**Be careful!** Do not try to cut the pencil yourself.) Attach the alligator clips to opposite ends of the pencil lead; gradually slide one clip toward the other. What happens?

Morse code was invented in 1840 by Samuel Morse, an American inventor and painter. Each letter of the alphabet is represented by a simple combination of long and short electrical signals that can easily be transmitted down a single wire. The code is written as dots, dashes, and spaces. Before the days of satellites and fax machines, all international newspaper reports and messages were sent flashing and buzzing down telegraph wires by Morse code operators.

MAKE it WORK!
Make your own Morse code tappers to send and receive secret messages.

You will need
2 pieces of wood	2 batteries
2 bulbs and bulb holders	wire
2 strips of copper	paper clips
2 slices of cork	glue and screws
hacksaw (and an adult to use it)	

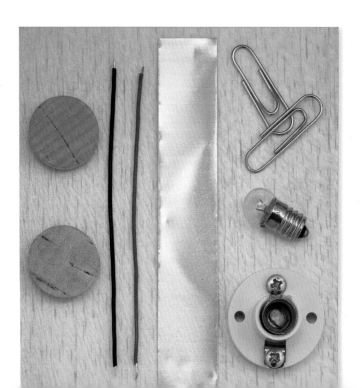

▼ International Morse code
These are the Morse code symbols. As you can see, they are made up of dots, dashes, and spaces. A dot is transmitted by pressing and instantly releasing the transmitter key. To send a dash, hold the key down twice as long as you did for the dot. A space between letters is the same length as a dot, and a space between words is the same length as a dash.

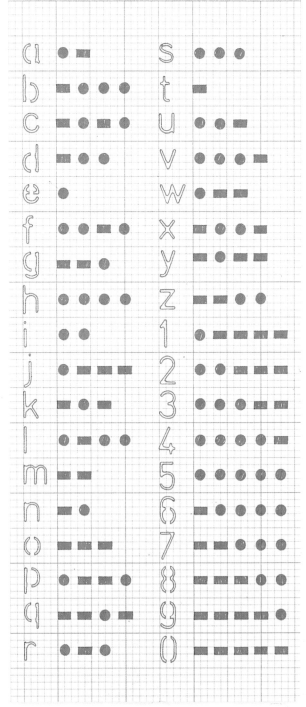

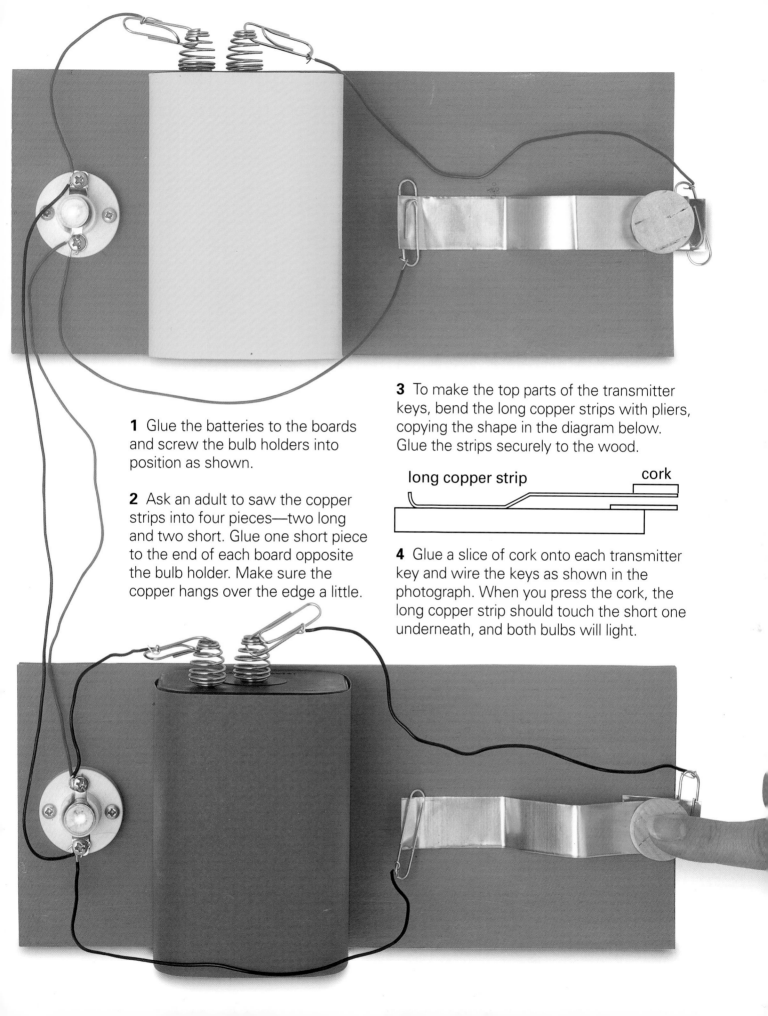

1 Glue the batteries to the boards and screw the bulb holders into position as shown.

2 Ask an adult to saw the copper strips into four pieces—two long and two short. Glue one short piece to the end of each board opposite the bulb holder. Make sure the copper hangs over the edge a little.

3 To make the top parts of the transmitter keys, bend the long copper strips with pliers, copying the shape in the diagram below. Glue the strips securely to the wood.

long copper strip   cork

4 Glue a slice of cork onto each transmitter key and wire the keys as shown in the photograph. When you press the cork, the long copper strip should touch the short one underneath, and both bulbs will light.

Some circuits are made up of lots of different connections that act together to perform complex tasks. Electrical equipment, such as radios, uses many tiny circuits, and the circuits themselves may not be made of wire but tiny strips of metal printed on a sheet. In computers, thousands of microscopic circuits are crammed onto one **silicon chip**.

MAKE it WORK!

Most circuits are made on a circuit board. The wires are spaced out so they cannot accidentally touch one another. This question-and-answer game shows you what a simple circuit board looks like. Each connection, when correctly made, will complete a circuit, and the bulb will light up.

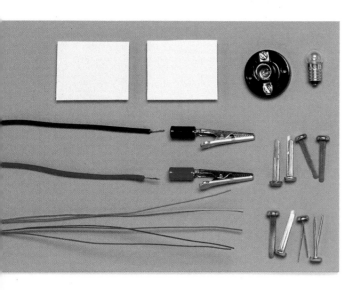

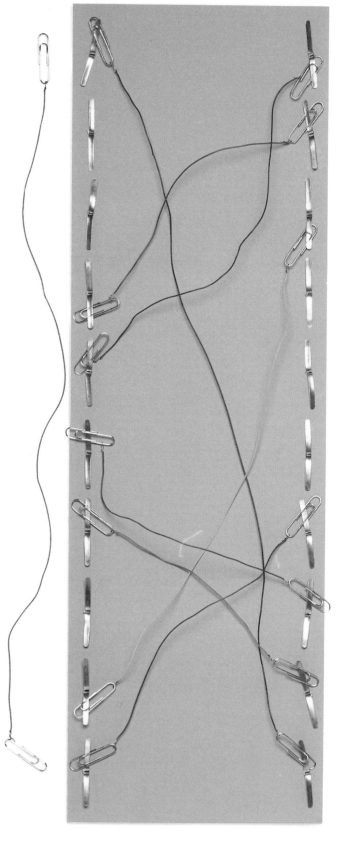

You will need

wire and battery
bulb and bulb holder
alligator/paper clips
stick-on hook-and-
   loop fasteners

stiff cardboard
paper fasteners
colored pens
magazine pictures
buzzer

1 Cut a piece of cardboard for your quiz board. Push in paper fasteners along each side of the cardboard. On the front of the cardboard, stick strips of hook-and-loop fasteners next to each paper fastener.

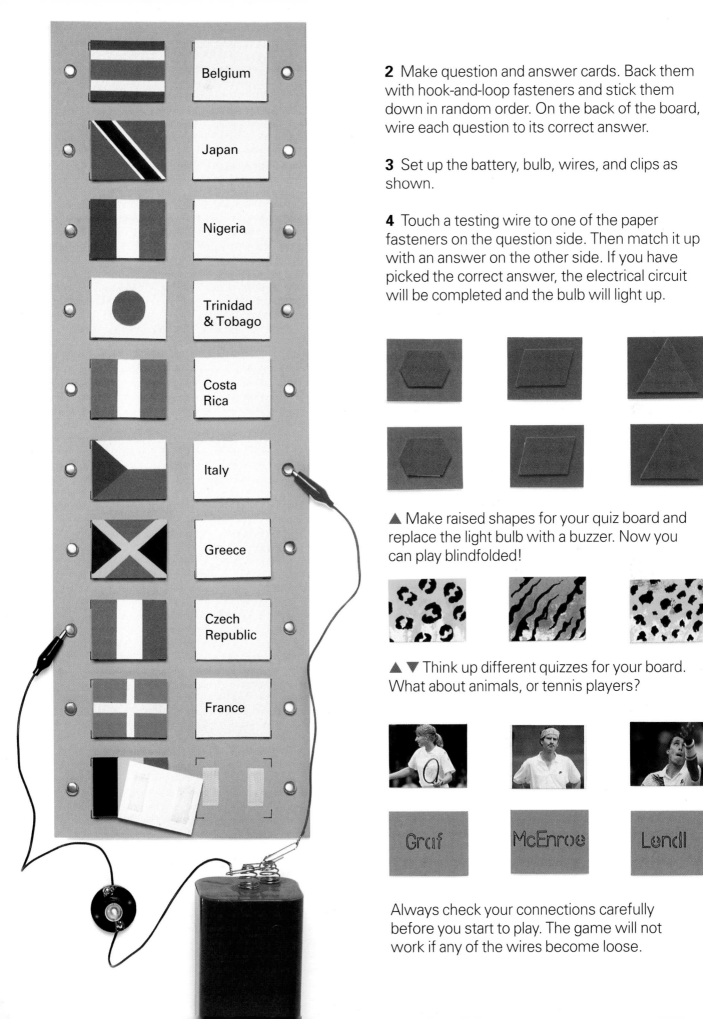

2 Make question and answer cards. Back them with hook-and-loop fasteners and stick them down in random order. On the back of the board, wire each question to its correct answer.

3 Set up the battery, bulb, wires, and clips as shown.

4 Touch a testing wire to one of the paper fasteners on the question side. Then match it up with an answer on the other side. If you have picked the correct answer, the electrical circuit will be completed and the bulb will light up.

▲ Make raised shapes for your quiz board and replace the light bulb with a buzzer. Now you can play blindfolded!

▲ ▼ Think up different quizzes for your board. What about animals, or tennis players?

Always check your connections carefully before you start to play. The game will not work if any of the wires become loose.

Like static electricity, **magnetism** is a natural, invisible force that makes certain objects attract or repel one another. The ancient Greeks noticed that certain rocks were magnets. Today we know that magnets work because of the motion of electric charges.

What is a magnet?

A magnet is a piece of iron or steel that attracts or repels certain other pieces of iron or steel. Like all substances, metals are made up of tiny particles called **molecules**, which in turn are made up of atoms. Normally, all the molecules in a piece of iron are facing in different directions. However, if we can rearrange the molecules so that they all face the same way, they will act together as a magnet, making a powerful force.

MAKE it WORK!

You can watch the power of magnets at work with this fishing game. The object is to catch as many high-scoring fish as possible. Players take turns fishing, and the winner is the player with the highest score.

1 Cut out large, medium, and small fish shapes from different colors of cardboard.

2 Draw in the eye, gills, and mouth with a black marker.

3 To get a fish-scale effect, tap paint onto the fish using a stencil brush and a piece of wire mesh. Use a lighter-colored paint for the belly. When the paint has dried, write a score number on each fish.

You will need

a small magnet with a hole in the center
thin cardboard   paint
black marker   utility knife
dowel and string   thin wire mesh
paper clips   stencil brushes

4 Make a fishing line by tying the magnet to the string. Attach the other end of the string to the dowel rod.

5 Attach a paper clip to each fish's mouth.

6 Make a sea from a cardboard box painted blue. Put in the fish and start fishing!

Around a magnet is an area called a **force field**, where the pull or push of metal and magnet is at its strongest. The force field may be strong enough to pass through wood or glass. The ends of a magnet, where most of the energy is directed, are much more powerful than the middle.

MAKE it WORK

The more powerful the magnet, the larger and stronger its force field. See how a small magnet works through cardboard and how the force field of a strong steel magnet can pass through a wooden door.

For the insects you will need

cardboard	glue
slices of cork	paint and paintbrushes
metal thumbtacks	horseshoe magnet
utility knife/ scissors	

Magnetic insects

Ask an adult to cut out insect shapes from cardboard with a utility knife. Paint the shapes and glue each one onto a small square of cork. Push a thumbtack into the other side of the cork. From the other side of a door, use a strong horseshoe magnet to make the insects move.

For the soccer game you will need

white cardboard	green cardboard
utility knife	cork
metal thumbtacks	dowels
small magnets	glue
tiny ball	paints or crayons

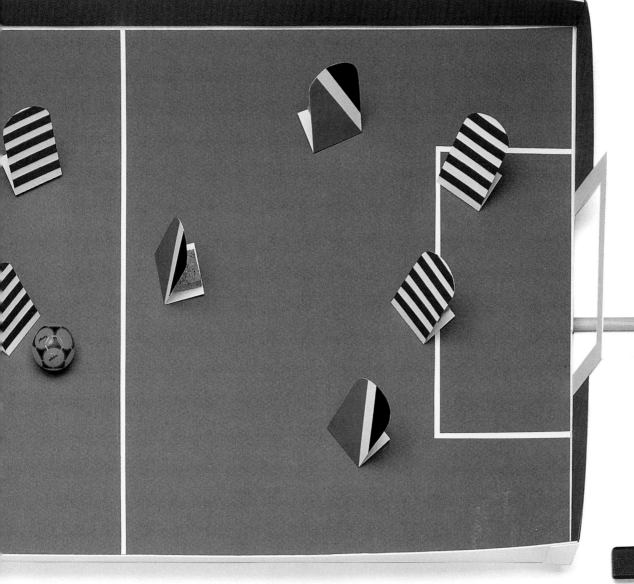

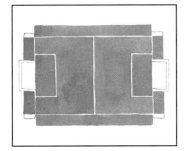

Table soccer

Make a box out of green cardboard, folding and gluing the corners as shown. Draw the lines of the soccer field in white. Make cardboard players with a piece of cork glued to the inside of each base. Stick a thumbtack through from the outside. The players are moved from under the box by magnets attached to dowels.

The force fields of magnets can pass through many different substances. The magnetic insects and magnetic soccer on the previous pages work because magnets can attract through wood and cardboard. A magnetic force field can also pass through water.

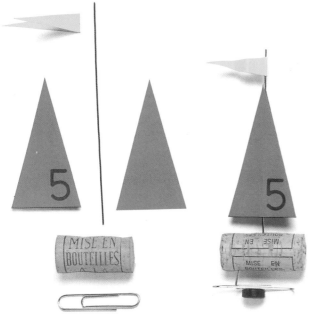

MAKE it WORK!

There are two different kinds of magnetic boats to make. The cork boats work by magnet-to-magnet attraction. The boat magnets are close to the bottom of the water container, so the boats can be pulled around by a small bar magnet attached to the end of a stick. The balsa-wood boat has thumbtacks pushed into its keel and needs a stronger magnet with a force field that will attract through shallow water.

Even though the ancient Greeks knew of magnets, for hundreds of years people did not know how to make magnets for themselves. Not until the nineteenth century were magnetism and its close connection to electricity properly understood.

You will need

colored poster board	wire
corks	paper clips
door magnets	waterproof glue
dowels	strong magnets
balsa wood	wooden skewer
metal thumbtacks	glass baking dish

To make the cork boats

1 Make the sails out of poster board. You can make one triangular sail by cutting out two triangles and sticking them back to back with a wire mast in the middle.

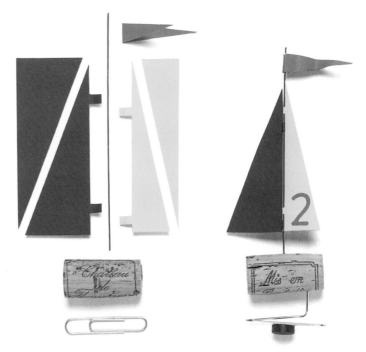

2 To make a more complex rig, cut a rectangle of poster board diagonally, leaving enough poster board on the long outside edge to make two tabs. Attach the sail to a piece of wire with these tabs.

3 Push the wire mast into a cork. **Be very careful!** Don't stab yourself with the wire. Top the mast with a flag made from a folded strip of paper of a contrasting color.

4 Unbend a paper clip as shown above. Push one end into the underside of the cork and glue a small door magnet onto the other end, using waterproof glue.

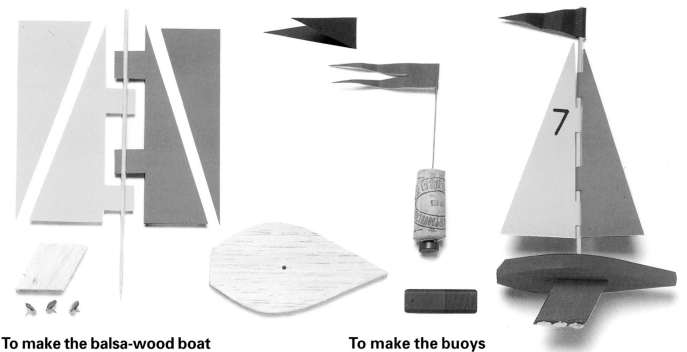

To make the balsa-wood boat

1 Ask an adult to help you cut a deck and keel out of balsa wood as shown above.

2 Stick a wooden skewer into the center of the deck to make a mast.

3 Make sails and flags as for the cork boats, but in a larger size.

4 Glue the boat together with waterproof glue and paint it. Push three thumbtacks into the bottom of the keel.

To make the buoys

Put a short piece of wire into a cork and top it with a colored flag. Stick a door magnet to the bottom of the cork with waterproof glue.

The ancient Greeks told a story about an imaginary island of magnetic mountains that could pull the iron nails out of passing ships!

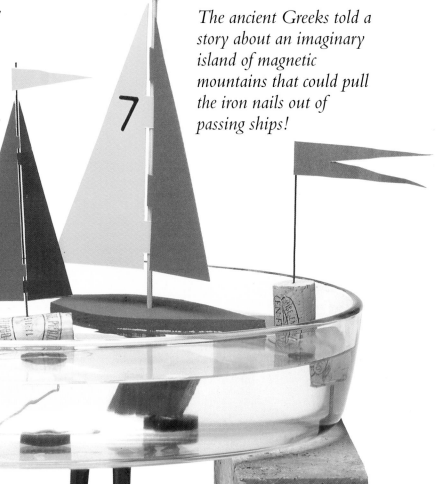

32 Magnetic Earth

Earth is actually a giant magnet, and its strongest points are near the North Pole and the South Pole. No one really knows why Earth is a magnet, but its force field extends thousands of miles into space. Any magnet on Earth allowed to swing freely will always point to the north—which is very handy if you need to find out where you are!

1 Cut out a circle of cardboard and cut a hole in the center just smaller than the opening of the yogurt container.

2 Using a protractor, divide the circle into accurate quarters and label the four compass points: North, South, East, and West.

3 Make the needle magnetic by stroking it with one end of a magnet about twenty times. Always stroke in the same direction. Tape the needle onto a thin slice of cork.

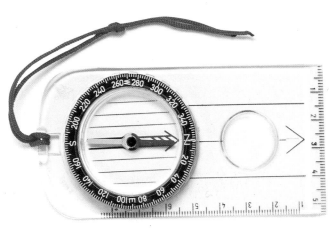

MAKE it WORK!

Every magnet has two points called **poles** where its magnetic force is strongest. Because magnets always line up with magnet Earth, the end of a magnet that points south is called the south pole. This is the basic principle that makes a compass work. Although some fancy compasses can give very precise readings, even a simple compass will let you know if you are heading in the right direction. All you need is a magnetized needle that can swing freely.

To make a water compass you will need

old yogurt container	magnet
needle	slice of cork
cardboard	protractor
tape	scissors

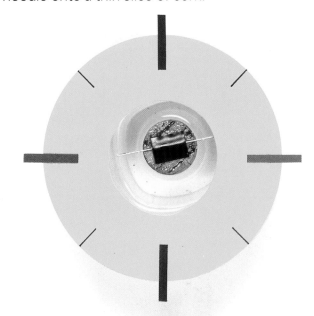

4 Fill the yogurt container with water and float the cork in it. When the needle has settled to the north, tape the cardboard ring to the container so that the needle points to the *North* label. Check your readings against a real compass.

The poles of magnets react to one another just like the two kinds of electric charge. Opposite poles attract—and like poles repel.

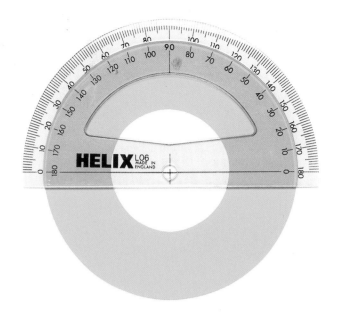

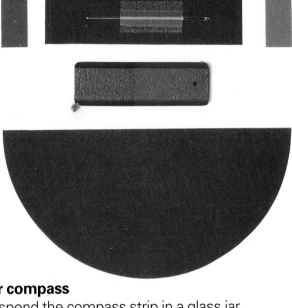

For two simple compasses you will need

cardboard   wooden skewer
magnet   needles and thread
jar   tape

Fold a strip of cardboard and tape a magnetized needle to it.

Jar compass

Suspend the compass strip in a glass jar, using a straw or a pencil and some thread. This compass will also work outdoors because the jar protects the needle from the wind.

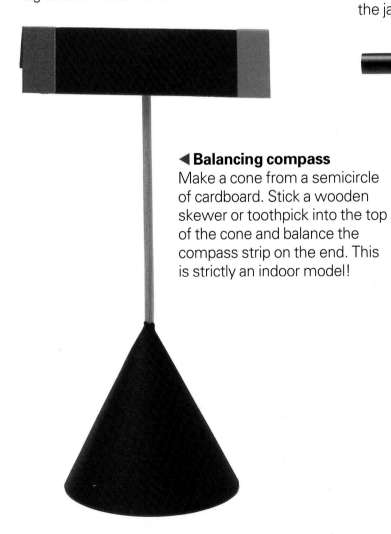

◀ Balancing compass

Make a cone from a semicircle of cardboard. Stick a wooden skewer or toothpick into the top of the cone and balance the compass strip on the end. This is strictly an indoor model!

You can actually see a magnetic field by laying a piece of paper over a magnet and sprinkling iron filings onto it. The magnet's force field is strong enough to work through the paper, and the filings will act like tiny magnets, clustering around the north and south poles where the force is strongest.

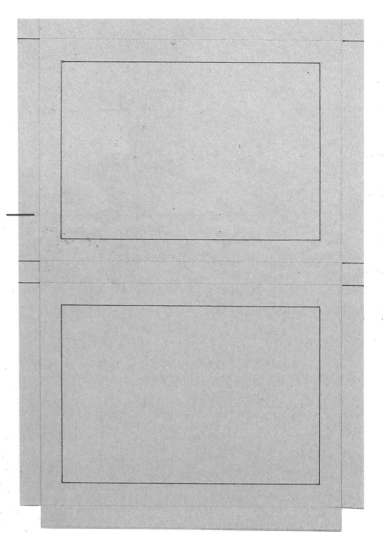

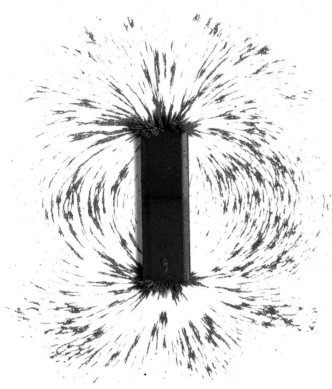

▲ Here you can see the force field at work. The iron filings form a pattern of lines running from pole to pole. These lines are called lines of induction, and they show us the invisible force field of the magnet.

MAKE it WORK!
Use the power of magnetism to draw pictures with iron filings.

You will need
cardboard	clear acetate sheets
rubber bands	iron filings and a magnet
scissors	clear tape

Be careful! Iron filings are dangerous. Don't breathe them in or swallow any, and don't lick your fingers after touching them.

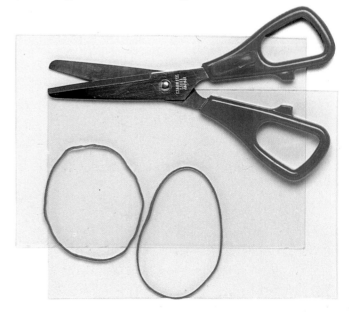

1 Cut out a rectangle of cardboard and mark it as shown. Cut along the red lines and fold and glue along the pencil lines to make a box.

2 Take two pieces of clear acetate and cut them slightly bigger than the windows of the drawing box. Tape them in place with clear tape.

3 Draw some silly faces on sheets of white paper, leaving out the hair.

4 Put a face card inside the box and shake iron filings on top of it. Snap the box shut with the rubber bands. Put the drawing box on a flat surface. Now you can "draw" the hair on your face by passing a magnet over the acetate.

*A new kind of train in Europe runs on the principle of magnetic levitation. Both the track and parts of the train are magnetic. The train works on the pull and push of magnets that repel and attract. The train floats above the track because the train and the track repel one another. There are no wheels to wear out or to cause **friction** and slow the train down.*

Electricity and magnetism are closely related to each other. Both are caused by the movement of electrons, and every electric current has its own magnetic field. This magnetic force in electricity can be used to make powerful **electromagnets** that can be turned on and off at the flick of a switch.

MAKE it WORK!

This crane uses an electromagnetic coil. The magnetic field produced by a single wire isn't very strong, but when electricity flows through a wire coiled around a conductor, the coil becomes a powerful magnet.

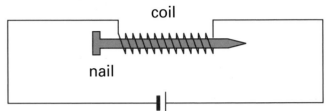

coil

nail

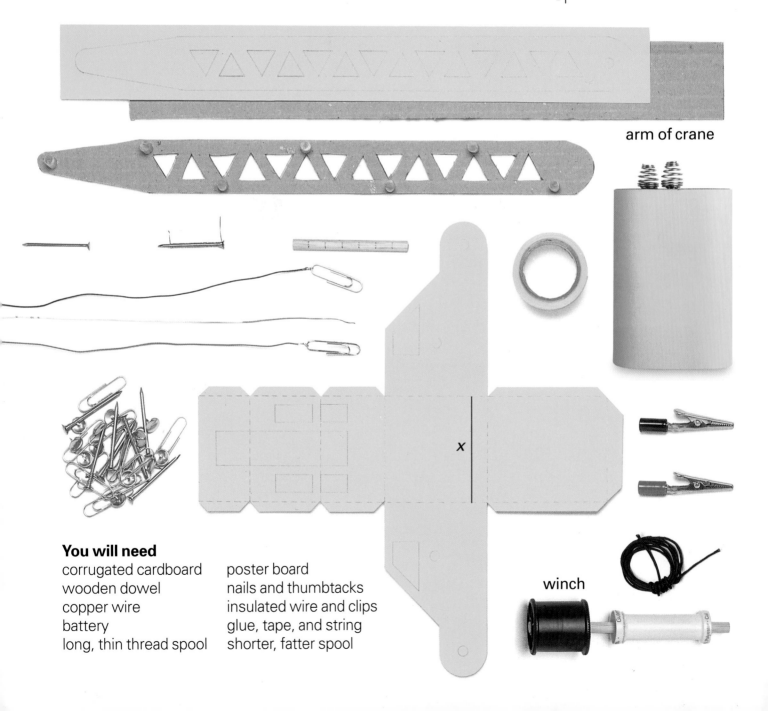

arm of crane

x

winch

You will need

corrugated cardboard	poster board
wooden dowel	nails and thumbtacks
copper wire	insulated wire and clips
battery	glue, tape, and string
long, thin thread spool	shorter, fatter spool

1 Draw the shapes shown on the opposite page and below on poster board and corrugated cardboard. The line marked *x* must be the same length as the long, thin spool.

2 Cut along the solid lines and fold along dotted lines to make the crane's body, arm, and base. Glue corrugated cardboard to them for support.

3 Assemble the winch as shown below, using the two spools, dowel, and string. Attach the other end of the string to the dowel at the tip of the crane arm.

4 Wrap copper wire around an iron nail to make an electromagnet. Connect it to a battery with insulated wire run over the top of the crane arm.

5 Check the circuit diagram to make sure the crane is correctly wired. When the wire and battery are connected, the nail will become magnetized and you can pick up a load of thumbtacks. Disconnect them, and the tacks will fall.

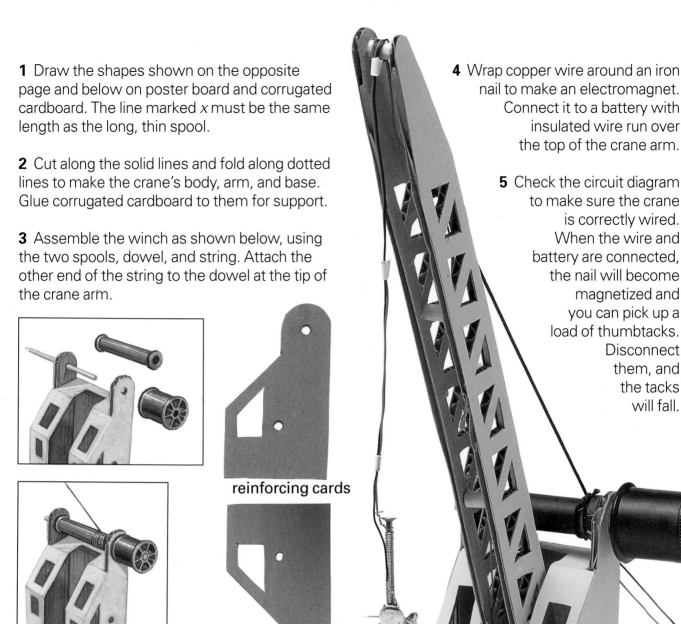

reinforcing cards

base support

body of crane

Electrical energy can be converted into mechanical energy, that is, energy that can pull and push and make things go. When electricity flows through the wires inside this motor, it makes them magnetic. The coil of wires becomes an electromagnet. It is attracted to fixed magnets inside the motor, which makes it spin around and around.

MAKE it WORK!

In this electric motor, a copper coil (the electromagnet) is connected to a battery by a clever little device called a **commutator**. The commutator brushes up against the wire ends of the electromagnetic coil, so that the electric current passes through, but the connections are loose enough to allow the coil to rotate freely.

As the coil turns, the connections of the commutator switch from side to side, so the direction of the electric current keeps changing. As the direction of the current changes, the poles of the electromagnet change sides, too. The electromagnet is always attracted to the farthest fixed magnet and so it keeps on spinning and spinning.

coil (electromagnet)

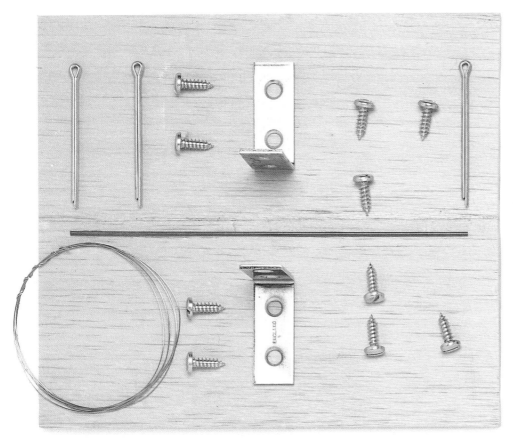

You will need

wood for the base	utility knife	copper wire	2 strong magnets
2 angle brackets	screws	balsa-wood block	electrical tape
thin copper tube	split-pin clips	alligator clips/paper clips	clear tape
	metal spindle	battery	

1 Cut the base board out of a piece of balsa wood, or find a piece of soft wood. You could paint it a bright color.

2 Ask an adult to drill a hole through the length of a small block of balsa wood. The hole should be wide enough for the copper tube to fit through.

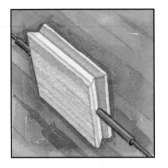 

3 Ask an adult to cut grooves along two edges of the balsa-wood block with a utility knife. Wrap copper wire tightly around the block. Insulate one end of the copper tube with clear tape. Fix the ends of the copper wire in place over the clear tape with insulating tape.

4 Screw the angle brackets to the base board. Stick on the magnets, positioning them so that they attract one another. Ask an adult to drill three holes along the center of the board to stand the split pins in.

5 Thread the metal spindle through the split pins and the balsa-wood block so that the coil is suspended and can turn easily.

6 Now make the commutator. The wires from the battery need to touch the ends of wire from the electromagnetic coil without stopping the coil from spinning. Strip some of the plastic from the ends of the wires and bend them inward. Follow the illustration on the left.

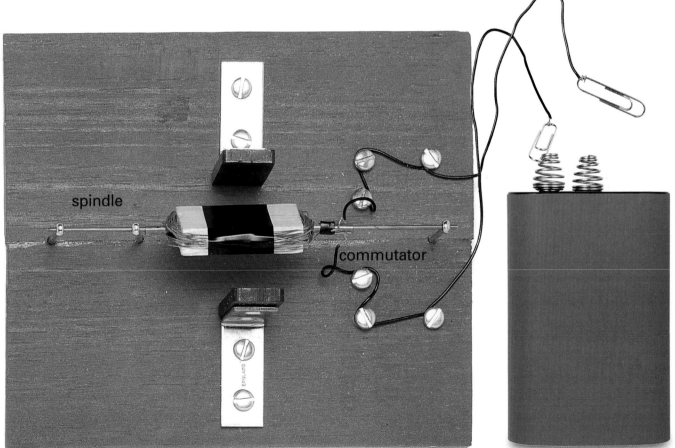

spindle

commutator

7 Screw the electrical wires into position so that the commutator connection is firmly fixed in place and cannot move. You may have to experiment a bit to get the screws positioned in the right place. Now connect the battery!

Our simple electric motor turned a spindle around and around. One of the most direct and efficient ways to use this energy is to attach a propeller to the spindle. Boats are often driven by propellers in this way.

MAKE it WORK!

This propeller-driven boat makes good use of energy. It is designed with a propeller that drives through air rather than water because air is thinner than water and easier to move.

You will need

6-volt battery
screws
balsa wood
thin dowels
electric motor
modeling clay
stiff metal wire
waterproof glue
model propeller
waterproof paint
long wires and clips
poster board and tape

To make the buoys

Push a couple of large nails into a cork. Add a flag and leave your buoy to float on the water.

1 Make a balsa-wood frame for the hull and deck as shown on the right. Ask an adult to drill holes for the dowels. Glue them in place with waterproof glue. Screw the electric motor in position. Paint the hull and deck with waterproof paint.

2 Connect wires to the motor.

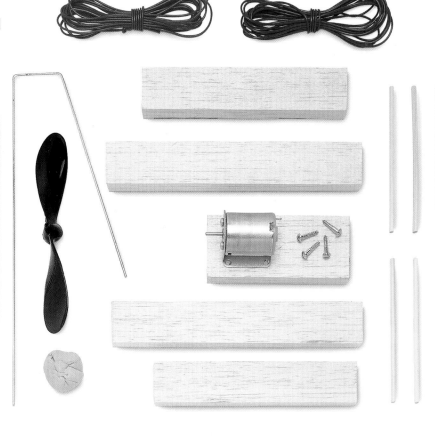

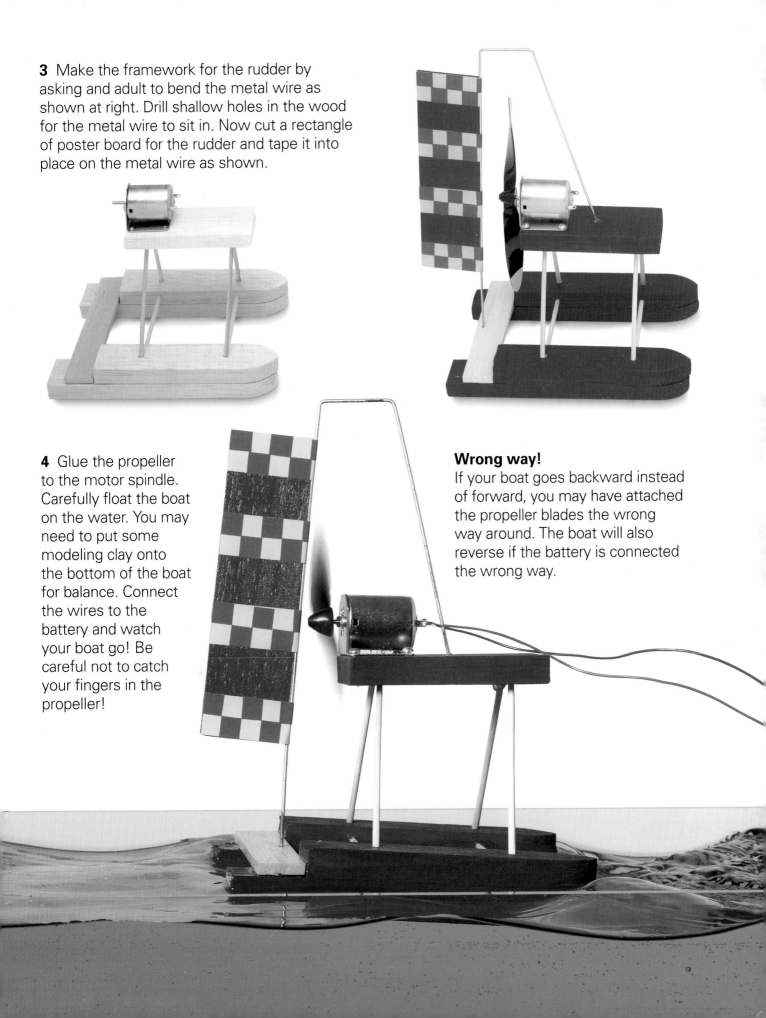

3 Make the framework for the rudder by asking and adult to bend the metal wire as shown at right. Drill shallow holes in the wood for the metal wire to sit in. Now cut a rectangle of poster board for the rudder and tape it into place on the metal wire as shown.

4 Glue the propeller to the motor spindle. Carefully float the boat on the water. You may need to put some modeling clay onto the bottom of the boat for balance. Connect the wires to the battery and watch your boat go! Be careful not to catch your fingers in the propeller!

Wrong way!
If your boat goes backward instead of forward, you may have attached the propeller blades the wrong way around. The boat will also reverse if the battery is connected the wrong way.

Modern high-speed electric trains are run by large electric motors. Power is supplied by overhead wires or electrified tracks. Electric trains use energy more efficiently because they don't have to haul heavy loads of fuel.

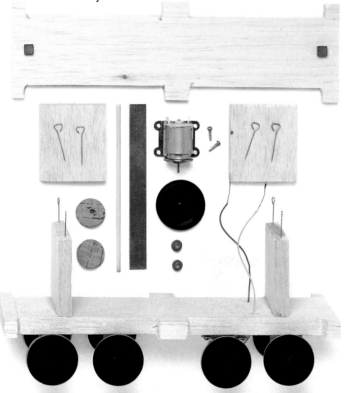

MAKE it WORK!
This model electric train works just like the real thing. The motor is on board, but the power

is supplied through power lines overhead and is passed down to the motor through the upholstery pins.

You will need
battery	beads
wooden dowels	paper clips
electric motor	slices of cork
poster board	upholstery pins
copper wire	3 thin copper strips
balsa wood	corrugated cardboard
7 plastic bottle tops	door magnets
screws, thin nails, and glue	

1 Ask an adult to help you cut out the balsa-wood base of the engine. Then glue and nail the two roof supports into position as shown. Stick the upholstery pins into the supports.

2 Glue a plastic bottle top to the spindle of the electric motor, making a wheel. Screw the motor to the base of the train, and then pass two wires from the motor up through the base and clip them to the upholstery pins.

3 Ask an adult to drill four holes into each copper strip and bend them as shown to make the axle holders. Drill holes through the base and screw the axle holders into place.

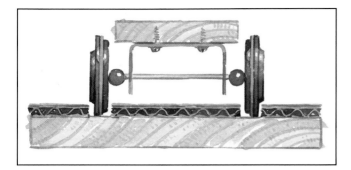

4 Make the axles by feeding the dowels through the axle holders. Then glue the beads and bottle tops to the ends of the axles to make the wheels.

6 Lay down three long strips of corrugated cardboard to make two grooves for your train to run along. Add balsa-wood arches overhead and string a length of wire between them, threading it through the upholstery pins. Connect the overhead wires to the battery and watch your train go!

Extra cars
Try making different kinds of cars for your train. They don't need motors or wires, but you can make wheels and bases as you did for the engine. Join them with small door magnets behind cork buffers.

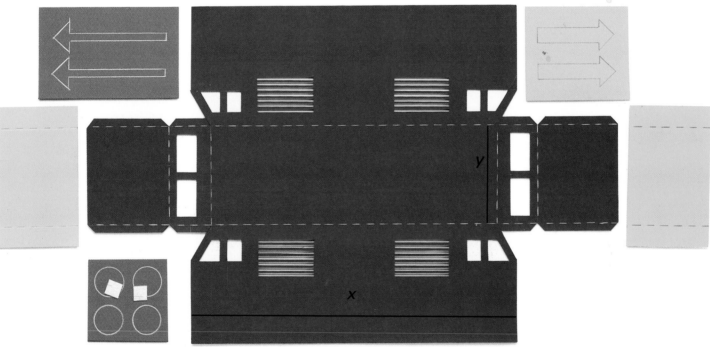

5 Draw a pattern on the poster board as shown, making sure that line *x* is as long as the balsa-wood base and line *y* is the same width as the base. Assemble the body of the train and fit it over the base.

Running backward and forward
To make your train run the opposite way, just reverse the connections on the battery.

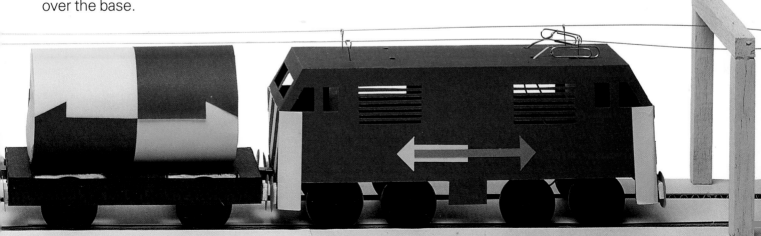

Electric motors range from the very small to the enormous. There are small battery–operated motors in model trains and clocks, but some electric motors in factories need a power supply so strong that it has to come directly from the power station.

MAKE it WORK!

See for yourself what happens if you change the supply to an electric motor. Build a spin-o-matico to make colorful patterns with paint. Compare the different results you get with different batteries.

1 Glue the two containers together, bottom to bottom, so you have two open ends. The top one will hold the paint, and the bottom one will hold the motor and wires.

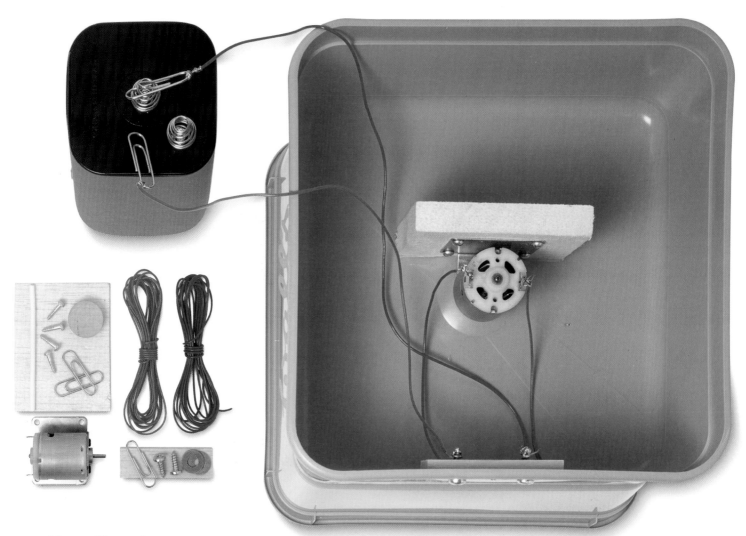

You will need

2 plastic containers	electric motor
2 different-sized batteries	paper clips
screws and washers	wire
slice of cork	wood
poster paint	glossy cardboard
glue and rubber cement	alligator clips

2 Poke the spindle of the electric motor through the center of the bottom container (the motor one), so it sticks up through the bottom of the top one (the paint holder). Screw the electric motor to a small piece of wood to hold it in place, and glue the wood to the container.

3 Make an on/off switch like the one shown on page 20, using a paper clip and two screws. Screw the paper clip to the outside of the container and onto a small piece of wood inside the container for support.

4 Wire the more powerful battery up to the motor, including the on/off switch in the circuit. To connect the wires to the switch, twist the ends of the wires around the screws, or attach them with alligator clips.

5 Turn the containers over. Put a slice of cork over the top of the electric spindle that is poking up through the container. Test the motor to see if the cork spins when you switch it on.

6 Take a piece of cardboard and stick it onto the cork using a dab of rubber cement.

7 Switch the motor on and dribble paint onto the whirling card to make a pattern. You can also use a paint brush if you want to.

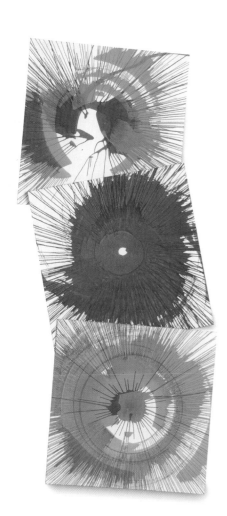

▲ Try doing some paintings using less power. Replace the powerful battery with the weaker one. How does the reduced power affect your finished painting?

When something is spinning very fast, the outside spins much faster than the inside. The force that seems to push everything toward the outside is called centrifugal force.

Acid One of two kinds of chemical substances. Acidic foods, such as lemons, taste sour or sharp. Strong acids are dangerous and can burn holes in wood or cloth.

Alkaline A word used to describe a chemical property of certain substances. In chemistry, an alkali is the opposite of an acid. Mixing things with acids and alkalis often causes chemical reactions, such as in a battery. Alkaline batteries are often used for bicycle lights, walkie-talkies, and other electric toys.

Atomic power Energy that comes from making changes to the center of an atom. By splitting atoms, an enormous amount of heat is created. This heat is used to boil water, making steam to drive turbines and produce electricity.

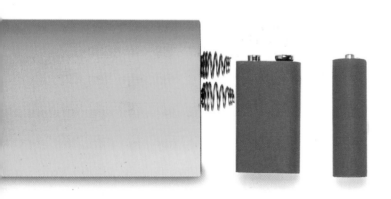

Atoms Tiny particles, over a million times smaller than the thickness of a human hair. Everything is made up of atoms—they're like building blocks, and by different atoms combining in different ways, different substances are created.

Circuit A loop-shaped path along which electricity can flow.

Commutator A special kind of electrical connection. It is used in electric motors. A commutator makes the direction of the electric current change at regular intervals.

Component In electronics, a component is one single part of a whole circuit. For example, a switch or a battery is a component in an electrical circuit.

Conductor In electronics, a conductor is any substance that an electric current can pass through.

Current electricity Current electricity is the electricity we use in homes, offices, and factories. It is produced in power stations and is then distributed around the country through wires, pylons, and transformers.

Electrolyte A liquid solution that is able to conduct electricity. Batteries use electrolytes to make electricity.

Electromagnets When an electric current passes through a metal, such as a piece of iron or copper, it always produces a magnetic field. Electromagnets are especially useful because their magnetism can be switched on and off with the electric current.

Electrons Tiny particles of atoms. Each electron carries an electrical charge.

Electroscope A scientific instrument used to measure the strength of an electrical charge.

Energy Energy is needed to do any job or action. Motors and engines use energy, and our bodies do, too. The food we eat and the electricity that powers an electric motor are called energy sources.

Experiments Special methods and procedures designed to test hypotheses about how the world works.

Filament A thin coil of wire, usually made from the element tungsten, inside a light bulb.

Force field The area around a source of energy (such as a magnet) where the energy works.

Friction Friction is the resistance that occurs when one object rubs against another. Friction produces heat and makes the two objects stick together (like tires gripping the road).

Generator A machine that turns heat or movement into an electrical current.

Graphite The substance from which pencil leads are made. Pencils used to contain actual lead until people discovered that graphite was better for writing with.

Hydroelectric power Energy produced by the movement of water through a generator.

Hypothesis A scientific guess based on known facts.

Insulators Materials that don't conduct electricity. Rubber and plastic are both good insulators.

Load The part of an electric circuit that uses the electric power. In a lighting circuit, the load is the light bulb.

Magnetism A natural, invisible force that makes certain metals attract or repel one another.

Matter All the different substances in the universe are matter. There are three forms: solids, liquids, and gases.

Molecule A tiny particle of a substance. Every molecule is made up of two or more atoms joined together.

Physics The study of energy and matter.

Pole One of two points on a magnet where the magnetic force is at its strongest. Like poles repel, and opposite poles attract.

Resistor A substance that offers resistance to an electric current.

Silicon chip A tiny, wafer-thin slice of the substance silicon that has an entire electronic circuit on it. Silicon chips are an important part of computers.

Static electricity An electrical charge, produced naturally when two things rub together. Lightning is the best-known example of static electricity.

Terminals In an electrical circuit, the points where the electric current leaves or enters the circuit.

Volt A unit that measures how much power is produced by a source of electricity.

Watt A unit that measures electrical power at the point where it is used in a circuit.

ancient Egyptians 13
ancient Greeks 26, 30, 31
argon 17
atomic power 4
atoms 6, 26

batteries 5, 10-11
battery tester 11
boosting power 44-45

centrifugal force 45
circuits 8-9, 12-19, 20-21, 24-25
coin battery 11
commutator 38, 39
compasses 32-33
computers 24
conductors 8, 9

Edison, Thomas 17
electric motors 5, 38-44
electric trains 42-43
electrodes 10
electromagnetism 36-37, 38-39
electrons 6, 7, 8, 9, 18
electroscope 6

filament 16, 17
fluorescent bulbs 17
force field 28, 30, 32, 34

generators 12
graphite 21

hydroelectricity 4, 12

insulators 8
iron filings 34

light bulbs 5, 16-17
lighthouses 12-13
lightning 7
lines of induction 34
load 9

magnetic fields *see* force field
magnets, magnetism 5, 26-35,
 38-39
molecules 26
Morse, Samuel 22
Morse code 21, 22-23

negative charge 6-7, 10
neon bulbs 17
nitrogen 17
nuclear fuel 12

parallel circuit 14, 15
Pharos of Alexandria 13
physics 4
poles (of the earth) 32
poles (of magnets) 32, 34, 38
positive charge 6-7, 10
power stations 12, 44

radios 24
resistance, resistors 21

series circuits 14, 15
silicon chips 24
static electricity 6-7, 8, 26
switches 20-21

terminals 8

Volta, Alessandro 11
volts 13

watts 13, 16